NECTAR AND POLLEN PLANTS OF UTAH

W. P. Nye

Research Entomologist
Federal Collaborator
Utah Agricultural Station

Monograph Series

Utah State University Press
Logan, Utah

Volume XVIII, Number 3
March 1971

SCHOLARLY PUBLICATIONS COMMITTEE

T. Y. Booth, *Chairman*
Charlotte Brennand
Calvin G. Clyde
LeGrand Ellis
Austin E. Fife
Allen D. LeBaron
John D. Schultz
James P. Shaver
John Mark Sorensen
Philip S. Spoerry
Mary Washington, *Editor*
Millard E. Wilde

Copyright, 1971, Utah State University

per copy price: $1.00

Requests for copies should be addressed to
Scholarly Publications Committee
Utah State University
Logan, Utah

INTRODUCTION

The first domestic bees were brought into Utah in the back of a covered wagon in 1848. Although the 1850 census listed only one hive of bees in Utah, by 1851 Brigham Young reported that several hives of bees were responding well to conditions in Utah. He requested pioneers to try beekeeping as a source for sweets and medicine. When the Deseret Bee Association was formed in 1872, Utah had approximately 2,000 colonies of bees. The 1880 law initiating bee inspection acknowledged that beekeeping was an important part of the food production and seed production program. It might be noted that the great seal of Utah displays a straw skep beehive, and that the original name for Utah was "Deseret," a *Book of Mormon* word denoting honeybee.

Considering Utah's desert environment, production of honey and wax has been extensive. Yet little has been published concerning improvement of beekeeping in the state.

One method beekeepers may use to improve production is to record the blooming period of the nectar and pollen plants in their vicinity. They may send unknown plants to the botany department of Utah State University for identification, and by using this monograph identify the common plants.

Oertell (1967) listed 182 species of these plants in the United States. Vansell (1931) listed 150 species of nectar and pollen plants in California, only six of which are principal sources of commercial honey production. He (1949) listed 90 species of nectar and pollen plants in Utah of which only alfalfa and sweet clover are sources of commercial honey. Wilson *et al.* (1958) listed 110 species of plants visited by bees in Colorado, and indicated that alfalfa and yellow sweet clover are the most important sources of honey.

The purpose of this monograph is to aid the beekeeper assemble data on the nectar and pollen plants in the vicinity of his apiary in order to manage bee colonies successfully. Such information enables him to calculate when he should install package bees, divide colonies for increase, requeen, use swarm control measures, remove honey, prepare colonies for winter, and locate profitable apiary sites. In addition, it tells him when to put on supers to provide the bees with plenty of room for storing nectar and pollen, and provide the queen with room for expanding the brood nest.

BEEKEEPING REGIONS

Based on flora, land topography, and beekeeping methods, Utah can be divided into four geographical regions: Wasatch Front, Uinta Basin, Delta (including Beaver), and Utah's Dixie (figure 1). These four beekeeping regions are designated by numbers in table 5 at the end of this publication. Number 1 is the Wasatch Front, 2 is the Uinta Basin, 3 is the Delta region, and 4 is the Dixie region.

The main sources of honey in the regions other than Dixie are alfalfa and the sweet clovers. In the Dixie area various species of native trees

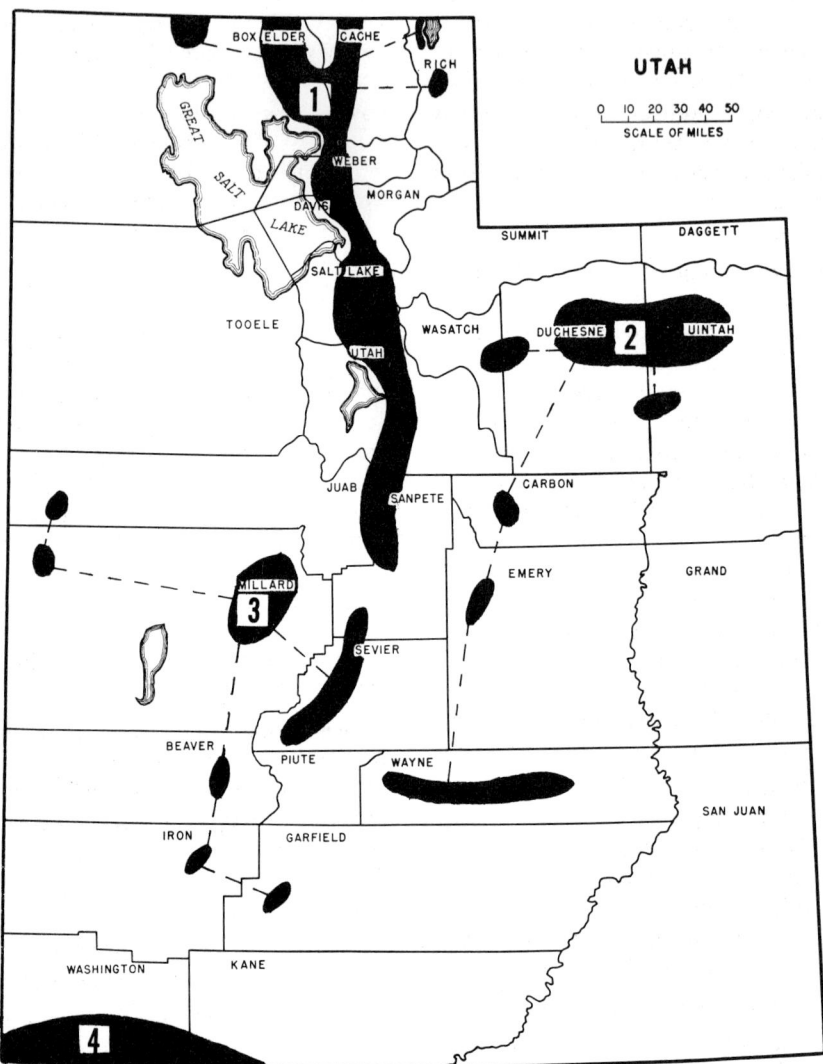

Figure 1. The four main honey-producing regions in Utah: 1 is the Wasatch Front, 2 is the Uinta Basin, 3 is the Delta region, 4 is the Dixie region.

INTRODUCTION

and brush plants such as mesquite, ceanothus, and arrowweed bloom at the beginning of the warm season and are important as early sources of honey. These four regions may also be divided by climate and topography. Colonies in the mountainous areas must be protected from the cold, and in certain forested areas from bears. In the desert, colonies must be protected from drifting sand.

The flora, climate, and topography determine the system of management practiced by the beekeeper. Migratory beekeeping is extensive in all the areas other than Dixie. But many colonies remain throughout the year in the Wasatch Front area and in the Uinta Basin, especially where fruit orchards, mustard, and miscellaneous early nectar and pollen sources are available for a spring buildup. Practically all honey bees are taken from the Delta region during the winter, partly because spring pollen and nectar sources are deficient. In Dixie, permanent apiaries may be established because of constant sources of pollen and nectar.

The northern area (Wasatch Front) comprises a narrow strip of territory west of the Wasatch Mountains which extends from Nephi northward to Idaho. It is an agricultural belt with deciduous fruit trees, and with alfalfa fields that are allowed to bloom at times. The northeastern area (Uinta Basin) includes agricultural sections of the land south of the Uinta Mountains. It is a colder area with a higher elevation than that of the Wasatch Front. Here beekeepers tend to pack colonies for winter with straw or roofing paper. West-central Utah (Delta area) comprises the productive valleys reaching approximately from Delta south to Beaver. This area is wind-swept during the winter, and most of the colonies are moved out for lack of fall and spring plants necessary for build up. Beekeepers in this area move their colonies to California or to the mountains on the eastern edge of Fillmore in Millard County.

Southwestern Utah (Utah's Dixie) includes the extreme southwestern portion of the state. It is lower in elevation than the other three regions, and colonies may be maintained year round. As pollen plants bloom in this area late in the fall, and as nut trees begin to bloom in January and February, bees have early and late sources of pollen and nectar. This area could produce queens and packing bees for shipment to the north.

Beekeepers, especially commercial operators, have learned that the nectar- and pollen-producing plants may change considerably over the years. Variations may be caused by droughts, changes in agricultural crops and practices, irrigation projects, and subdivision development. Modern agriculture depends greatly on the honey bee to fulfill its pollination needs. Honey bees are the most efficient and only manageable pollinators that man can use to increase a large number of food materials, because they visit flowers to collect nectar and pollen, and they do not destroy the plant in the pollination process. Although various species of bees contribute to the pollination of Utah crops, an estimated 80 percent of this pollination is done by honey bees.

While most of the surplus honey produced in Utah is from a few introduced crops grown under irrigation, some native and miscellaneous sources of nectar and pollen also are worthy of mention. Pollen sources are widely distributed, regardless of irrigation. Much of Utah is covered with small perennial bushes, including sagebrush, saltbush, winter fat, and greasewood, most of which are of practically no value to bees as sources of pollen except during limited periods.

Productive locations for honey production will become more difficult to find. If a housing development or new subdivision forces a beekeeper to move his colonies from his home, he may have difficulty finding a desirable location nearby. Except during certain years of abnormally heavy precipitation, honey bees would actually starve to death on much of the rangeland.

POISONOUS PLANTS

Fortunately the beekeeper seldom needs to be concerned about plants that are poisonous to honey bees. Locations with abundant growth of deathcamas (*Zigadenus* spp.) and locoweed (*Astragalus* spp.), some of which are poisonous to honey bees, should be avoided while in bloom. Damage to colonies from poisonous nectar and pollen may be severe in some years, but usually is of small consequence.

NECTAR SECRETION

Hobbyists in beekeeping frequently ask: "What plants can I grow that will increase my yield of honey?" It is generally not economically practical to grow a crop for the bees alone. Beekeepers are dependent on cultivated crops, such as alfalfa, grown for other purposes or on plants on which few honey bees work.

The quantity of nectar produced by different plants is extremely variable. The orange blossom is a copious producer; frequently these blossoms will fill with nectar to overflowing. On the other hand, the quantity produced by an alfalfa blossom is very small (0.19 ul per flower). The quantity present influences the rate of its collection by the nectar-gathering bees. The sugar concentration is a factor of bee visitation, and relative humidity at any given time is also.

Some nectars are colorless and others have a green, yellow, or brown tinge. Not only is the color variable but the amount of sugar present in the nectars also shows a wide range, a factor which undoubtedly has a great influence on bee activity. Bees evidently select, other things being equal, the nectars of high sugar concentrations. For ex-

NECTAR SECRETION

ample, sweet cherry blossoms are greatly preferred to those of the sour cherry. Sometimes a special kind of honey is stored, showing that bees continue to work certain nectar sources in preference to others.

Through the use of the sugar refractometer, it is possible to determine rapidly the sugar concentration (total solids) in nectars. A large number of nectars have been examined in this manner. Some data indicating the extreme difference in sugar concentrations appear in table 1.

Bees utilize nectar as their source of carbohydrate food. The nectar is sucked into the honey sac within the body and is thus carried into the hive. The thin nectar, which is mostly sucrose, is evaporated and inverted principally to dextrose and fructose in the comb cells to about 82 percent sugar, after which it is sealed over with beeswax. This thickened and elaborated nectar is the honey of commerce.

Nectar secretion or production is affected by such environmental factors as soil type, soil conditon, altitude, latitude, length of day, light conditions, and weather. Soil conditions, such as fertility, moisture, alkalinity, and acidity not only affect the growth of the plant but also the secretion of nectar. Luxuriant plant growth, for example, alfalfa grown for hay, does not necessarily imply that maximum nectar secretion will take place. A limited growth results in increased nectar production. Clear, warm, windless days are likely to favor nectar secretion.

Nectar is secreted by an area of special cells in the flowers called a nectary. Certain species, such as vetch, cotton, partridgepea, and cowpeas, produce nectar from tiny specialized areas in the leaves or stems called extrafloral nectaries.

POLLEN

Pollen is the fertilizing powder contained in the flower anthers. Fertilization is the process by which pollen causes the embryo to be formed. Pollen is essential to the nutrition of honey bees and the many species of wild bees that assist in the pollination of alfalfa and other plants. It contains most of the food elements required for the growth of young bee larvae. Though the chief food of adult bees is honey, both pollen and honey are necessary for larval bees. Not only is pollen used by the honey bees during the summer season, but ample pollen stores for winter are also important. When sufficient food is available in the hive, honey bees rear brood in the spring long before pollen is found in the field. Just how much pollen is required to maintain a colony in good condition is not known, but through the use of pollen traps as much as 50 pounds per colony was trapped from the incoming bees during one season at Logan, Utah.

Pollen from the blossoms is collected on the bee's hairy mouthparts, legs, and coat. It is combed from the hairs and formed into a ball on each of the bee's hind legs. One group of wild bees, including the leaf cutter species, transports the

NECTAR AND POLLEN PLANTS OF UTAH

Table 1. Average sugar concentration of some samples of nectar from various plants

Plant	Place of collection	Percent Sugar concentration
Alfalfa (**Medicago**)	California, Nevada, Oregon, Utah	41
Apple (**Malus**)	Hood River, Oregon; Logan, Utah	50
Apricot (**Prunus**)	Davis, California; Logan, Utah	15
Blackberry (**Rubus**)		
Himalaya	Western Oregon	27
Evergreen	Western Oregon	36
Cherry (**Prunus**)		
Sweet	Corvallis, Oregon; Logan, Utah	51
Sour	Corvallis, Oregon; Logan, Utah	20
Cotton (**Gossypium**) Acala:		
Blossoms	Fresno, California	22
Extra-floral nectaries	Fresno, California	41
Houndstongue (**Cynoglossum**)	Richmond, Utah	48
Orange (**Citrus**)	Orange, California	25
Pear (**Pyrus**), Bartlett	Medford, Oregon; Ogden, Utah	15
Red filaree (**Erodium**)	California, Oregon, Utah	65
Russian olive (**Elaeagnus**)	Logan, Utah	40
Sage (**Salvia**) Black	Southern California	45
Sweet clover (**Melilotus**):		
Yellow	Eastern Oregon; Logan, Utah	52
White	Eastern Oregon; Logan, Utah	35
Yellow star-thistle (**Centaurea**)	Davis, California	38

POLLEN

pollen in a hair brush under the abdomen. During the collection of a pollen load, the honey bee usually visits only one kind of plant, which accounts for the fact that the pollen grains in each load are practically all alike. Many of the wild bees, however, collect pollen from several sources on each field trip.

Many plants provide pollen for honey bees, but the quantity available and its collectability vary greatly from one source to another. The blossoms of deciduous fruits are good pollen sources and in some places, while they are in bloom, they constitute the chief source. The deciduous-fruit belt, lying west of the Wasatch Mountains, is undoubtedly an important pollen area during spring. Other important pollen sources are dandelion, balsamroot, mule ears, little sunflower, alfalfa, sweet clover, mustards, gum plant, oak and maple, orchard morning-glory, Canada and other thistles, corn, greasewood, poverty weed, and desert mallow. Russian-thistle, annual sunflower, cattail, elderberry, bassia, sugar beet, wild lettuce, and chicory are of minor importance.

None of the wild bees gather pollen from as many sources as do honey bees. Percival (1947) concluded that any plant offering a fair amount of pollen per flower-form will be worked for pollen by the honey bee. Many species of bees are limited to a few families of plants, and others are practically host specific.

Pollen grains from different plants vary physically and chemically. They range in size from about 15 to 160 microns. They may be spherical, triangular, square, disk-like, or crescent-shaped. Some pollen grains are provided with wings for their transportation by the wind. Others are light and apparently full of gases, which facilitate their transfer by air movement. The surface texture of pollen grains is almost infinite in its variety. The color, as seen in bee loads, ranges from almost white through all colors of the spectrum to almost black. The color of the anther pollen and the change of color effected when the pollen is collected by the honey bee is shown in table 2.

As the pollen of different plants more or less closely related is often very similar or even identical in appearance, it is frequently impossible to carry identification to the species level, and for this reason, pollen of each type is named after the principal family or genus represented. Many pollen types are rather dry (unexpanded) when the anther splits open, the living grain is relatively small, the outer wall is contracted, and the furrows appear as narrow slits. When the grain is moistened, it expands and the thin membranes of the furrows bulge out (expanded). The descriptive data for a few types are therefore arranged alphabetically by type, with the exception of the leguminous type, which, because of its importance, is considered first:

NECTAR AND POLLEN PLANTS OF UTAH

Table 2. The color of the anther pollen of the pollen types and the change of color effected when the pollen is collected by the honey bee

Pollen type	Anther pollen	Bee-collected pollen
Leguminous	Deep chrome yellow	Caledonian brown (W)
Red clover	9 E 2 (cream)	Khaki
Alfalfa	Lemon chrome yellow	Auburn and hazel (W)
Sweet clover	Lemon chrome yellow	Olive brown and citrine (W)
Basswood or linden		
Cactus	Golden yellow	Golden yellow and khaki (W)
Composite	9 L 3 (Sunflower Y.)	Yellow ochre
Aster	Sunflower yellow	Bronze yellow
Balsamroot	Deep chrome yellow	Terra cotta and raw umber (W)
Dandelion		
Coniferous	Ivory yellow	Olive brown (W)
Pine	Lemon chrome yellow	Olive brown (W)
Cruciferous—mustard	Lemon chrome yellow	Brass
Grass—sweet corn	White	Dark grey (W)
Labiate—catnip	White	Slate grey (W)
Rosaceous		
Apple	Lemon chrome yellow	Citrine (W)
Umbelliferous		
Carrot	Chinese yellow	Aztec

POLLEN

Leguminous type

Because of the importance of leguminous plants in the production of honey, the pollen has been quite thoroughly studied. The typical shape is ellipsoidal when seen from the side in the unexpanded form. Each grain has three openings arranged at equal intervals around the equator, and usually indicated by rather large protuberances in the expanded form. The grains are light colored, with very finely granular contents, and thin usually smooth walls. Longitudinal grooves are often seen as faint lines running in either direction from the openings.

Red clover (*Trifolium pratense* L.) pollen is of the typical shape, but has an average size of 38 µ to 45 µ. The walls are rather thick and have a slightly reticulate appearance (figure 2).

Alfalfa (lucerne) (*Medicago sativa* L.) pollen differs from red clover mainly in having a smoother and thinner wall and has an average size of 38 µ to 43 µ. The unexpanded forms are elliptical, with the ends regularly rounded. The expanded forms are nearly circular (figure 3).

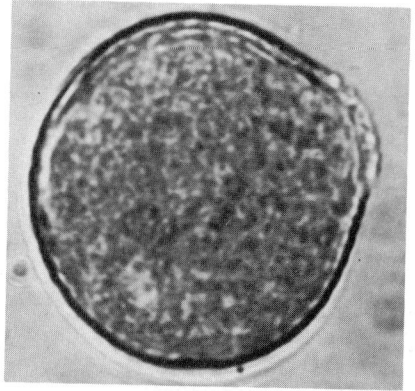

Figure 2. Red clover, expanded form.

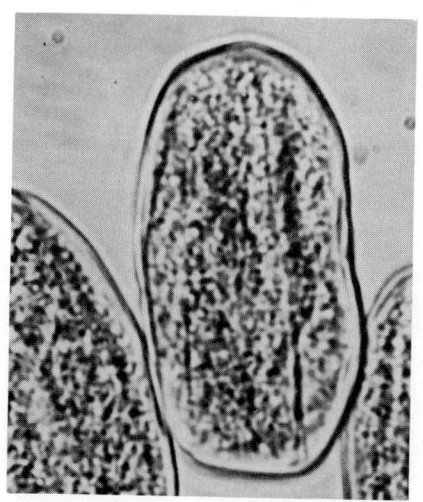
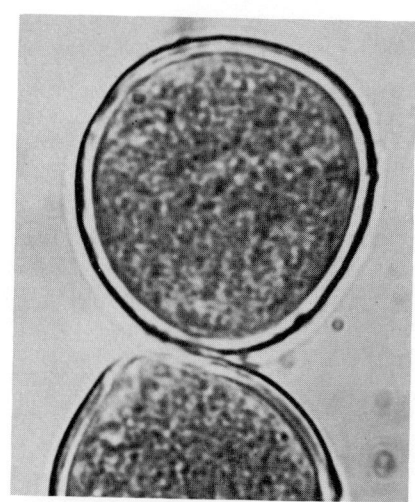

Figure 3. Alfalfa, unexpanded left and expanded right.

Sweet clover (*Melilotus spp.*) pollens can scarcely be distinguished from that of white clover. It is, however, a trifle longer in proportion to the diameter, averaging about 30 μ x 24 μ (figure 4).

Basewood Type

Basswood or linden (*Tilia* spp.) pollen is quite distinctive. The shape is somewhat lenticular, so as nearly always to present the same aspect. The appearance is that of a regular three-lobed figure inscribed in a circle. The three openings are subtended by clear spaces between the lobes. The diameter averages 32 μ (figure 5).

Cactus Type

Cactus (mostly *Opuntia* spp.) pollen grains are 125 μ in diameter of approximately spherical shape. They have a number of faces, each provided with an opening, and the intervening space is reticulated (figure 6).

Composite Type

There is but little difference in the pollen of the various species of Composita. It is usually rather small, spherical, provided with three openings, and covered with spines.

Aster spp. pollen is thickly covered with short spines. The grains are 30 μ to 35 μ in diameter (figure 7).

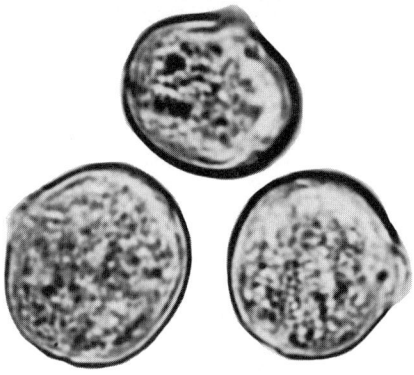

Figure 4.   Sweet clover.

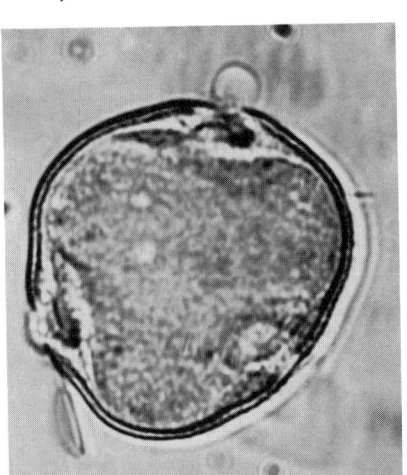

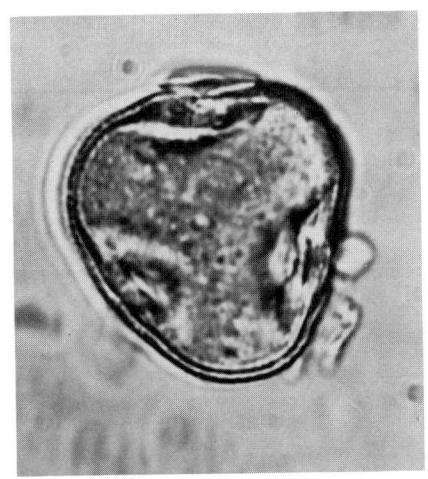

Figure 5.   Basswood or linden.

POLLEN

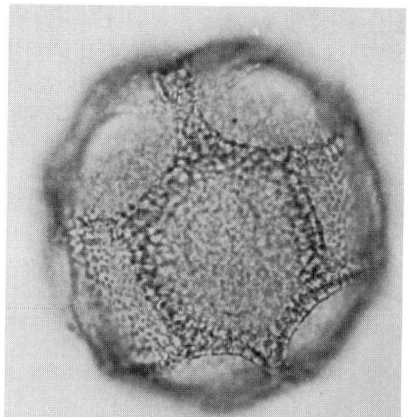

Figure 6. Cactus.

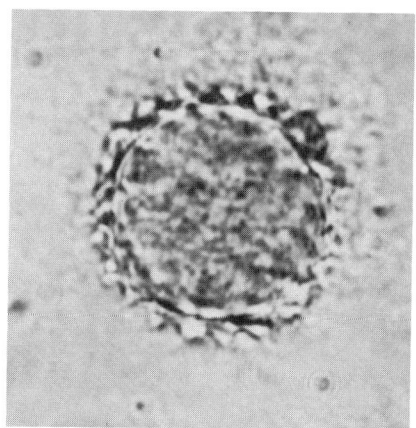

Figure 7. Aster.

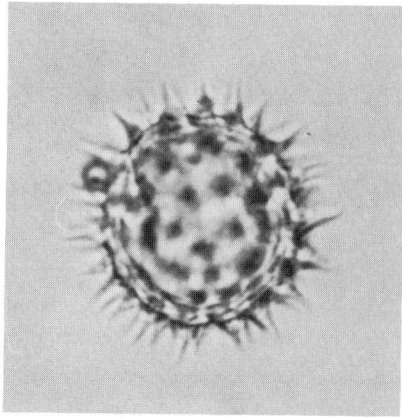

Figure 8. Balsamroot.

Balsamroot and mule ears (*Balsamorhiza sagittata* and *Wyethia amplexicaulis*) pollen is 35 µ to 40 µ in diameter and rather thinly covered with sharp spines (figure 8).

The chicoriaceous subtype pollen grains are polyhedral rather than spherical, with an opening on each face. The spines are small.

Dandelion (*Taraxacum officinale* Weber) pollen is typical of this sub- type and has a diameter of 35 µ to 40 µ. The grains are yellow (figure 9). *Cichorium intybus* L. pollen grains are of the same size and shape as those of dandelion except that they are white instead of yellow.

Coniferous type

Pine pollen consists of two dark bodies or floats connected by a curved sac constituting the pollen grain proper, which is clear and light colored (figure 10).

Cruciferous type

Mustard (*Brassica* spp.) pollen grains are dark yellow colored and spherical, with a finely indented surface and three openings situated in smooth areas. The diameter is about 32 µ (figure 11).

Evening primrose type

Evening primoses (*Oenothera* spp.) have large triangular pollen with roundish protuberances at the angles. The grains are 128 to 135

— 11 —

µ in diameter, with protuberances 40 µ or more in diameter (figure 12).

Grass type

The grasses have ovoidal pollen which is frequently almost spherical. The contents are often denser in the larger end. The surface is very smooth, and there is a single opening at the larger end, sometimes centrally placed, but usually located more or less to one side. Orchard grass has pollen 32 µ long, timothy pollen is 36 µ long, and in corn the average length is about 100 µ (figure 13).

Labiate type

The pollen grains of the mints have the shape of flattened ellipsoids; that is, they have three axes

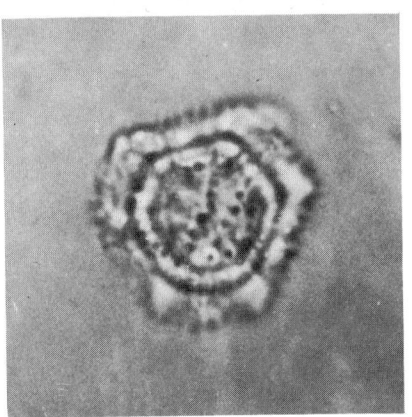

Figure 9. Dandelion.

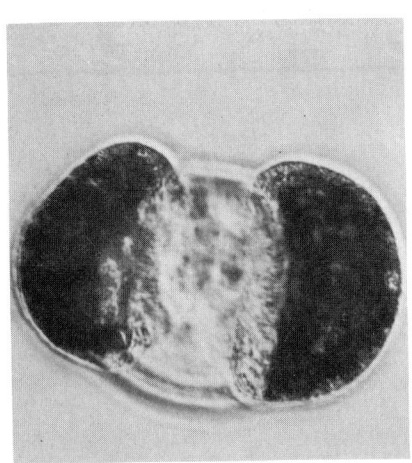

Figure 10. Pine.

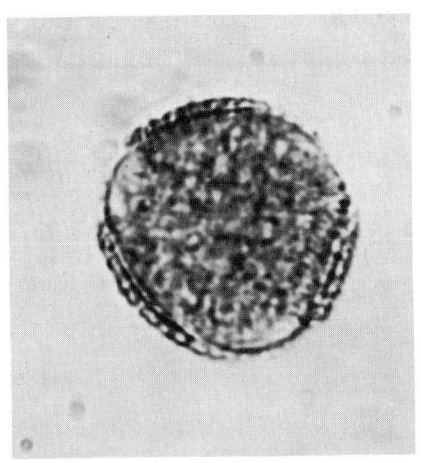

Figure 11. Mustard.

POLLEN

of different length. When viewed in the direction of the shortest axis, the shape is that of an ellipse, an elongated hexagon, or an ellipse with six indentations corresponding in position to the six openings, which are sometimes indicated by very small protuberances. These openings are arranged at equal intervals, two of them marking the ends of the medium diameter. When viewed from the side, the grain appears marked by grooves passing between the poles of the shortest diameter. The pollen of the different species is so much alike as to be in most cases

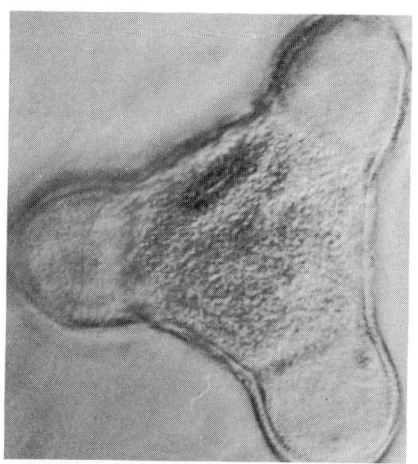

Figure 12. Evening primrose.

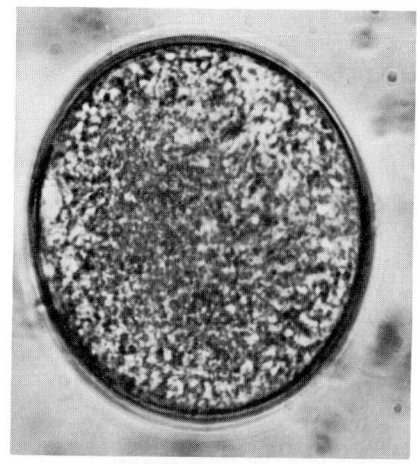

Figure 13. Corn.

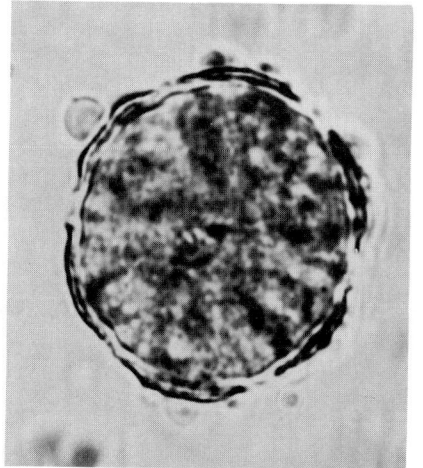

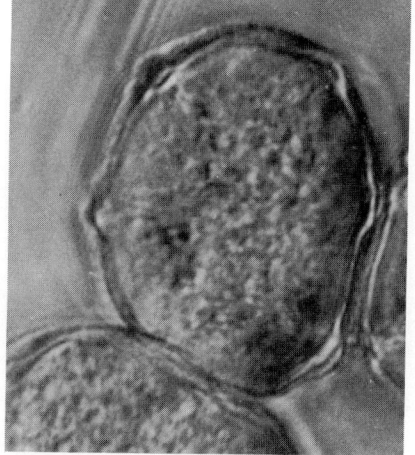

Figure 14. Catnip.

NECTAR AND POLLEN PLANTS OF UTAH

indistinguishable. Pollen grains of horehound (*Marrubium vulgare* L.), horse mint [*Agastache urticifolia* (Benth.) Kuntze], spearmint (*Mentha spicata* L.) and catnip (*Nepeta cataria* L.) have a length of 34 μ to 36 μ long (figure 14).

Mallow type

Plants of the mallow family have large spherical pollen grains covered with sharp spines which are proportionately rather short. Cheeses (*Malva neglecta* Wallr.), wild hollyhock [*Iliamna rivularis* (Dougl.) Greene], scarlet globemallow [*Sphaeralcea munroana* (Dougl.) Spach.] have similar pollen grains which are 70 μ or more in diameter (figure 15).

Rosaceous type

In general, the pollen grains are subtriangular with openings at the corners which sometimes appear to be covered with a shallow cap. Protuberances from these openings are usually cleft. Apple (*Malus* sp.) pollen has a diameter of 36 μ to 44 μ (figure 16).

Umbelliferous type

Wild carrot (*Daucus carota* L.) has pollen similar to that of clover but more narrow in proportion; size, 16 μ by 24 μ to 28 μ (figure 17).

RESULTS OF CHEMICAL ANALYSIS

Samples of pollen from 32 different sources, as reported by Todd and

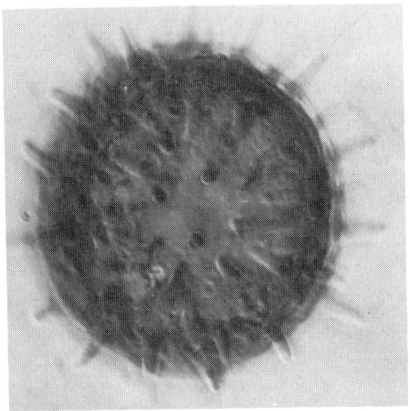

Figure 15. Cheeses.

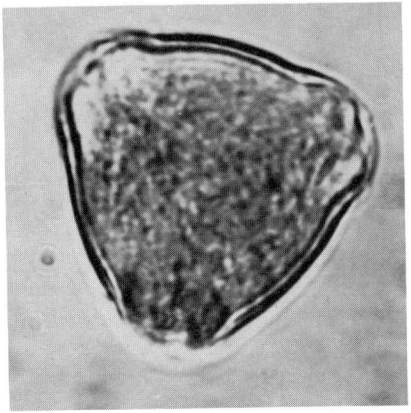

Figure 16. Apple.

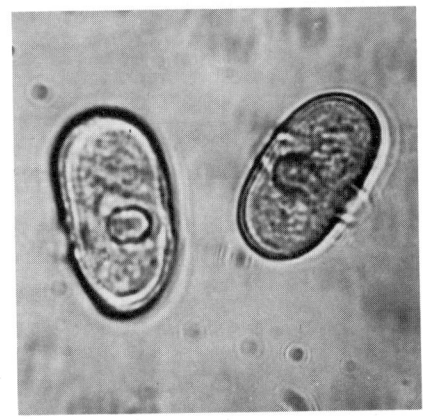

Figure 17. Carrot.

CHEMICAL ANALYSIS

Table 3. Summary of chemical analysis of 32 samples of pollen from different plants*

Constituent	Maximum Percent	Minimum Percent	Average Percent
Protein	35.50	7.02	21.38
Ether extract	17.55	0.94	4.97
Carbohydrates	48.35	1.20	28.63
Water	17.14	3.91	10.91
Ash	6.36	0.91	2.85
Undetermined	57.23	21.65	31.25

* Todd, Frank E., and Ormond Bretherick. The composition of pollens. J. Econ. Ent. 35:312-317. 1942.

Table 4. Amino acid content of average pollen and sweet corn pollen, expressed as percent of crude protein

Component	Average pollen	Sweet corn pollen
Arginine	5.3	4.7
Histidine	2.5	1.5
Isoleucine	5.1	4.7
Leucine	7.1	5.6
Lysine	6.4	5.7
Methionine	1.9	1.7
Phenylalanine	4.1	3.5
Theronine	4.1	4.6
Tryptophane	1.4	1.6
Valine	5.8	6.0

Brentherick (1942) are summarized in table 3.

More recently Standifer (1967) reported that bee-collected pollens are comparatively rich in carbohydrates. Reducing sugars range from 15 to 43 percent, with an average of about 29 percent. The glucose, fructose, sucrose, raffinose, and stachyose content is not significant,

NECTAR AND POLLEN PLANTS OF UTAH

although the bees apparently utilize those that are available. Corn pollen is high in starch. The pollen shell is not utilized by bees, but is eliminated with the feces after the internal matter has been removed by digestive processes.

Standifer reported the protein value of pollen varies from 10 to 36 percent. The amino acid content of average pollen and sweet corn pollen with a crude protein of 26.3 and 26.9 percent, respectively, is shown in table 4.

The amino acids, except threomine, are essential for normal growth of the young adult bee. With the exception of histidine and perhaps arginine, they cannot be synthesized by bees and must be obtained from the pollens consumed.

Other constituents of pollen reported by Standifer (1967) are as follows:

Constituents	Amount (Percent)
Fats	1.3 - 19.7
Minerals (ash):	
Calcium	1.0 - 15.0
Chlorine	.6 - .9
Copper	.05 - .08
Iron	.01 - 12.0
Magnesium	1.0 - 12.0
Phosphorus	.5 - 21.6
Potassium	20.0 - 45.0
Silicon	2.0 - 10.4
Sulfur	.8 - 1.6
	(Micrograms per gram identifed)
Vitamins:	
Ascorbic acid	131.0 - 721.0
Biotin	.19 - .73
D	.2 - .5
E	0 - .32
Folic acid	3.4 - 6.8
Inosital	.3 - 31.3
Nicotinic acid	37.4 - 107.7
Pantothenic acid	3.8 - 28.7
Pyridoxine	2.8 - 9.7
Riboflavin	4.7 - 17.1
Thiamine	1.1 - 11.6

Undoubtedly, the chemical constituents of pollen have a bearing on its value to bees, and this factor may in some way influence their selection of pollens.

NECTAR AND POLLEN PLANTS

Honey bees have been observed visiting many Utah plants for nectar and pollen. Of these, only the most important are discussed at any length. However, an annotated list of both major and minor sources appears in table 5 at the end of this publication. The blossoming periods for some plants, as reported annually by Utah beekeepers in regions 1, 2, and 3, are given in figure 18. The solid line indicates full bloom and presumably the main yielding dates. The exact time of blossoming is somewhat variable, depending on season, elevation, and exposure.

Alfalfa

Alfalfa is the chief honey plant of Utah, and in many locations it is the only source of surplus honey (figure 19). The plant is grown in practically every location where water is available and the salt content of the soil is low. It does not thrive, how-

PLANTS

Plant	Feb.	Mar.	Apr.	May	June	July	Aug.	Sept.	Oct.
Willow			-- — --						
Fruit bloom			-- — — --						
Dandelion			-- — —						
White clover					-- —		-- —		
Alsike clover					-- —	— —			
Alfalfa (first cutting)					-- —	--			
Alfalfa (second cutting)						— -- —	—		
White sweetclover					-- — —	— —	—		-- —
Yellow sweetclover					-- —	— --			

Figure 18. Plant blossoming dates reported by Utah beekeepers.

ever, in very wet places. The total area in this crop is about a half million acres. This makes about 10 acres to each colony of bees. Utah Common is the most extensively used alfalfa variety, but Grimm, Ladak, Atlantic, Buffalo, Vernal, Uinta, Ranger, and others are also grown. Two or three cuttings of alfalfa hay are general throughout the state. The last crop is sometimes harvested by livestock. As a rule, spring opens late and blossoming does not begin in most areas and seasons until toward the end of May. Since the control of the alfalfa weevil is necessary, the first crop is seldom allowed to blossom. The second crop, blossoming primarily late in July and in August, provides most of the bee pasturage.

The Uinta Basin, the Delta and Cedar City areas, and parts of Box Elder, Cache, Utah, and Juab Counties produce large amounts of alfalfa hay. Seed is also produced in these and other areas. In general, the crop left for seed constitutes the most important source of honey, since blossoming is then at its maximum. In some areas bees collect alfalfa pollen readily while in others they do not.

Alfalfa is largely self sterile and must be fertilized by pollen from another alfalfa plant. The anthers and stigma (the male and female parts) form a sexual column that is tightly enclosed by the petals. This column is suddenly released (or tripped) and strikes the bee on its head, leaving behind a mass of sticky pollen. When the bee visits another alfalfa blossom, the process is repeated and pollination results. The untripped and tripped alfalfa blossom are shown in figure 20.

A honey bee bred to cross pollinate alfalfa with a load of pollen on her legs, indicating it has pollinated hundreds of blossoms, is shown in figure 21.

Nectar concentration in alfalfa blossoms varies considerably. A

NECTAR AND POLLEN PLANTS OF UTAH

range of less than 30 percent sugar to over 60 percent has been recorded from refractometer samples obtained from honey bees' stomachs. Almost invariably, when bees are working a field with wet soil, the nectar concentrations are low, but the opposite takes place on relatively dry soils. Other factors may also enter into this difference in sugar concentration; for instance, the influence of variety has not been established.

Figure 19. Alfalfa in full bloom.

PLANTS

Figure 20. Alfalfa flower, untripped on left and tripped on right.

NECTAR AND POLLEN PLANTS OF UTAH

Honey obtained from the alfalfa plant is almost white. The nectar in the honey stomach of the bee appears to be colorless. Honey high in dextrose, as is that from alfalfa, tends to granulate rapidly. Extracted alfalfa honey granulates with a fine-grained texture, and for this reason is highly desirable for sale as a granulated product. It has a mild flavor and lacks the vanilla-like flavor, so characteristic of sweet clover honey. The honey production of Utah is estimated at about 2.5 million pounds per year, and 75 percent of that which comes on the market is estimated to be from alfalfa.

Aster

A conspicuous fall flower. It is a source of some nectar and pollen. One of the last fall sources.

Figure 21. Honey bee, bred to cross pollinate alfalfa, with a load of pollen on her legs indicating that she has pollinated hundreds of blosoms.

PLANTS

Astragalus

The number of *Astragalus* species in Utah is exceptionally large and, when climatic conditions are favorable, some of them are abundant. Bees work some of these plants freely for nectar and pollen, especially when a more attractive source is not available. The term "loco" is applied to those that are poisonous to livestock. Serious bee poisoning has been caused by spotted loco (*Astragalus lentiginosus* Dougl.) in the vicinity of Beaver. Just how much influence the presence of loco plants has had on the total bee losses in Utah is not established.

Balsamroot and mule ears

There are two genera of these plants [*Balsamorhiza* (figure 22) and *Wyethia*], but in their appearance and attractiveness to bees they are so similar that they are discussed as one. Numerous and extensive patches of these plants grow on the higher foothills and in the lower mountain areas, particularly in cen-

Figure 22. Balsamroot, an important early summer source of pollen and nectar.

NECTAR AND POLLEN PLANTS OF UTAH

Figure 23. Bitter brush.

PLANTS

tral and northern Utah. The large yellow blossoms provide both nectar and pollen, which are attractive to bees. Large pollen reserves from balsamroot are seen frequently in colonies that have been within reach of it. The pollen is a rich orange color. The nectar as seen in the stomachs is bright yellow. In many places both oak trees and balsamroot grow within flight range of an apiary, and the combination of the two plants should provide fine supplies to build up the colonies for later flows. Balsamroot begins to blossom late in the spring and continues for 4 to 6 weeks. On the steep slopes there is at least 10 days difference between the blossoming time on the lower elevations and at 1,000 feet above. These plants are not palatable to livestock.

Bitter brush

Bitter brush [*Purshia tridentata* (Pursh.) DC.] is an aromatic, much branched shrub 2 to 8 feet tall, often

Figure 24.   Bitter brush blossom.

NECTAR AND POLLEN PLANTS OF UTAH

associated with sagebrush in our foothills and mountains. Its numerous yellow blossoms, ½ inch in diameter, are freely visited by bees for pollen and nectar. It blooms during May and June (figures 23 and 24).

Black locust

Black locust (*Robinia pseudoacacia* L.) is widely planted about city and country homes. The tree thrives under many soil and climatic conditions. The white blossoms hum with bees, all of which seem to collect nectar only. It is also important to bumble bees. The bloom on a particular tree lasts only a short time, but there is considerable variation in the blossoming time of different trees.

Boxelder

Boxelder (*Acer negundo* L.) is an early spring source of pollen. The blossoms are inconspicuous (figure 25).

Broom snakeweed

Broom snakeweed is also called matchweed and small rabbitbrush. The low-growing bushy plants grow in numerous places, sometimes abundantly. Yellow blossoms appear late in summer and bees visit them eagerly for pollen and nectar. The

Figure 25. Bees visit boxelder blossoms for pollen and possibly nectar.

PLANTS

pollen is a golden-orange color and easily obtained by bees.

Buttercup
(Ranunculus jovis A. Nels)

The sage buttercup is an early flowering succulent plant that sends up leaves and waxy blossoms early in the spring while the ground is still wet from melting snow (figure 26). The bright yellow blossoms are proterandrous. The stigmas are mature before the anthers of the innermost stamens have dehisced. Bees visiting the flowers for nectar and pollen become covered with pollen, alight on the middle of the flower, and crossing must result.

Cactus

Cactus is associated with drier regions in Utah. In years of good rainfall, it flowers from April to June on blow sand and gravelly washes where it provides both pollen and nectar (figure 27).

Cane fruits

All the bush berries are sources of nectar and pollen for a short pe-

Figure 26. Buttercup.

riod. Raspberries bloom in June and July and are a leading honey plant in some areas where fruit is produced. The honey is white with a delicious flavor. Raspberries are visited freely by bees for nectar and pollen.

Catalpa
(Catalpa speciosa Warder)

This large tree with large leaves and long pods has great clusters of white flowers spotted with yellow and purple-brown (figures 28 and 29). The flowers are fragrant, producing much nectar. They are visited by smaller Hymenoptera and by bees which crawl into the large bell-shaped corolla. The stamens are bilobed and sensitive. The honey is thin and watery and it is rarely obtained pure. Catalpa blooms in June.

Cleome

The purple blossoms of the Rocky Mountain bee plant are conspicuous. Large patches of this plant occur in favorable places almost throughout the state. Sometimes it gives surplus honey (figure 30). A yellow-blossomed species of the same genus

Figure 27. Cactus.

PLANTS

Figure 28. Catalpa tree.

NECTAR AND POLLEN PLANTS OF UTAH

common in some areas of Utah is less important for honey, but it is a good source of pollen. It was probably the principal host of the alkali bee before plants such as alfalfa, sweet clover, and bindweed (orchard morning glory) were introduced.

Clovers

Clovers, including sweet (figure 31), white, red, and alsike, are, next to alfalfa, the most important source of honey in Utah. Both the yellow and white sweet clovers are found in numerous places. Fields for seed or

Figure 29. Catalpa blossom.

PLANTS

for livestock pasture are seen in various valleys in the northern part of the state. However, the greater part of the sweet clover is of miscellaneous distribution and is found along roadsides, railroad tracks, irrigation ditches, and wherever sufficient water is available to mature the plants. Flowers appear in June and continue until frost, but the main flush of bloom is in midsummer.

Some farmers do not plant sweet clover, because of its tendency to persist after the fields are turned to other crops. It is objectionable in alfalfa seed fields, because the seeds contaminate the alfalfa seed and are difficult to separate out. In wheat fields the green stalks give difficulty at harvest time, particularly when combine machines are used.

While it has not been possible to obtain nectar from the blossoms by the pipette method, evidence from nectar taken by bees indicates that soil moisture affects the sugar concentration. For example, the sugar concentration of nectar in the honey sacs of bees that worked plants on wet soil was 33 percent or less, while that of nectar originating in plants on dryer land often exceeded 55 percent.

White sweet clover has a better reputation among beekeepers as a honey producer than does yellow, notwithstanding the fact that yellow sweet clover provides richer nectar. Presumably the quantity per plant is less in the yellow variety.

Figure 30. Cleome, called Rocky Mountain bee plant, where abundant, is a source of surplus honey.

NECTAR AND POLLEN PLANTS OF UTAH

Both sweet clovers provide pollen in abundance. Judging from the size of bee loads and quantities appearing in the combs, however, bees gather more from the yellow. Almost invariably loads are large when bees are working yellow sweet clover, but in white sweet clover loads tend to be small or medium in size, and it apparently requires considerable time for a bee to obtain a load. Almost without exception, whenever bees are working sweet clover, numerous pollen gatherers are present. Apparently bees cannot obtain nectar from white sweet clover without becoming daubed up with pollen, which they frequently brush off for storage in the pollen basket. This tendency is also apparent when bees are working white clover, where the blossoms are small. Yellow sweet clover blossoms are much larger than those of white sweet clover.

A considerable amount of sweet clover honey originates in the Uinta Basin, in Cache Valley, in the upper end of the Salt Lake Valley, and other smaller areas, particularly in isolated mountain valleys of regions 1 and 2. The honey from both types of sweet clover has a characteristic vanilla-like flavor, which is very acceptable to the trade. It granulates with medium rapidity into a fine-grained mass.

White or Dutch clover is used in permanent livestock pasture. It se-

Figure 31. White sweet clover, one of the major sources of surplus honey.

PLANTS

cretes nectar freely but because much water (that is, continuously wet top soil) is required to keep it growing, the acreage in the state is limited. The honey is colorless and slow to granulate. Ladino clover is a giant variety of white clover.

Alsike clover, a moisture lover, is grown to a limited extent, and wherever found, it contributes to the honey crop. Honey from alsike clover is nearly colorless and of pleasing flavor.

Honey bees collect a considerable amount of brown-colored pollen from red clover, but as a source of honey, this clover is disappointing. Bumble bees, however, prefer red clover for both pollen and nectar.

Cottonwood

The cottonwood (Populus supp.) trees, of which there are several species, are among the earliest sources of pollen. Many homes throughout Utah are surrounded or flanked with stately Lombardy poplar. The catkins are easily worked and productive. Bees frequently continue to collect from them after they have dropped to the ground. The brownish-orange-colored bee loads of pollen are soft and oily.

Cucumber

The cucumber (*Cucumis sativus* L.) is widely cultivated and is dependent upon bees for pollination of its flowers, since the pistillate and staminate flowers are in separate flowers of the same plant, that is monoecious. Honey bees visiting the flowers for nectar and pollen become covered with pollen (figure 32). Only where large fields of cucumber are cultivated is any appreciable amount of honey obtained. The honey is pale yellow amber, strong at first but later becoming mild.

Dandelion

A plant familiar to all. It provides an attractive and dependable supply of pollen and nectar for bees. The honey is bright yellow and granulates quickly into a solid that looks like butter (figures 33 and 34).

Deciduous fruits

The primary deciduous fruits grown are apple, apricot, cherry, peach (figures 35 and 36), pear, and plum. All are important early pollen sources for a few weeks each spring. Fruit blossoms are disappointing for honey production, but each kind provides some nectar. The chief value of fruit blossoms is in stimulating greater brood production, to recoup the colony popula-

Figure 32. Cucumber.

NECTAR AND POLLEN PLANTS OF UTAH

Figure 33.   Dandelion, produces much orange pollen and yellow nectar.

PLANTS

tion after the natural winter shrinkage.

Blossoms of some fruit varieties, for example, sweet cherry, are reported as wholly self-sterile; that is, each tree requires insect-borne pollen from another variety of the same species. With the almond, sweet cherry, and apple, cases of inter-sterility between varieties are commonly reported. Honey bee colonies are easily placed in an orchard to assist in pollen distribution (figure 37).

Figure 38 shows a honey bee collecting pollen from an apple blossom.

Some bee losses follow the application of pesticide sprays to apple

Figure 34. Honey bee collecting nectar from dandelion flower.

— 33 —

NECTAR AND POLLEN PLANTS OF UTAH

Figure 35. Peach, a good source of pollen and some nectar in the spring.

PLANTS

Figure 36. Close-up of peach blossom.

NECTAR AND POLLEN PLANTS OF UTAH

Figure 37. Many kinds of deciduous fruit blossoms must be cross-pollinated by insects or by hand. Beehives in the orchard provide suitable insects for pollination.

PLANTS

Figure 38. Honey bee collecting pollen from apple blossoms.

NECTAR AND POLLEN PLANTS OF UTAH

and pear trees. The use of a summer-flowering cover crop in these orchards is particularly hazardous because of blossom contamination by dripping spray.

Elm
These trees blossom and mature seed in early spring, from the last of March to the middle of April, long before the leaves appear. They provide a liberal supply of grayish-colored pollen for bees (figure 39).

Evening primrose
Evening primrose (*Oenothera caespitosa* Nutt.) grows close to the ground with basal leaves and stemless flowers on long floral tubes (figure 40). The flowers are perfect symmetrical nearly 4 inches across, snow white, turning to pink with age and fragrant (figure 41). It grows on dry sandy hillsides at lower and middle elevations. It blooms from May to

Figure 39. Chinese elm.

Figure 40. Evening primrose.

— 38 —

PLANTS

July. The nectar is secreted by the ovary, where it is protected from rain and yet is easily accessible to insects. The blossoms are visited by honey bees, bumble bees, solitary bees, butterflies and moths. There are a number of solitary bees that are host specific to evening primrose for both nectar and pollen (figure 42).

Figwort
(Scrophularia lanceolata Pursh)

Plant is open, erect, branching and often 5 feet tall. The blossoms are brown and pale green with maroon markings at the base (figure 43). It is usually confined to rich woods in open areas and forms thick stands on stream banks in our canyons. The nectar is secreted very abundantly by two glands at base of the flower. Figwort is visited so extensively by wasps that it is often referred to as a wasp flower. Honey bees and wasps, which largely pollinate the flowers, approach from the front and come in contact with the stigma, leaving some of the pollen on it. The following day the stamens come in view and the style becomes flabby and rests on the lower lip. An insect in going to the flower for nectar is dusted with pollen from the later flowers. Self-pollination cannot occur. Honey bees collect nectar

Figure 41. Blossom of of evening primose.

NECTAR AND POLLEN PLANTS OF UTAH

Figure 42. Wild bee collecting pollen from evening primrose blossom.

PLANTS

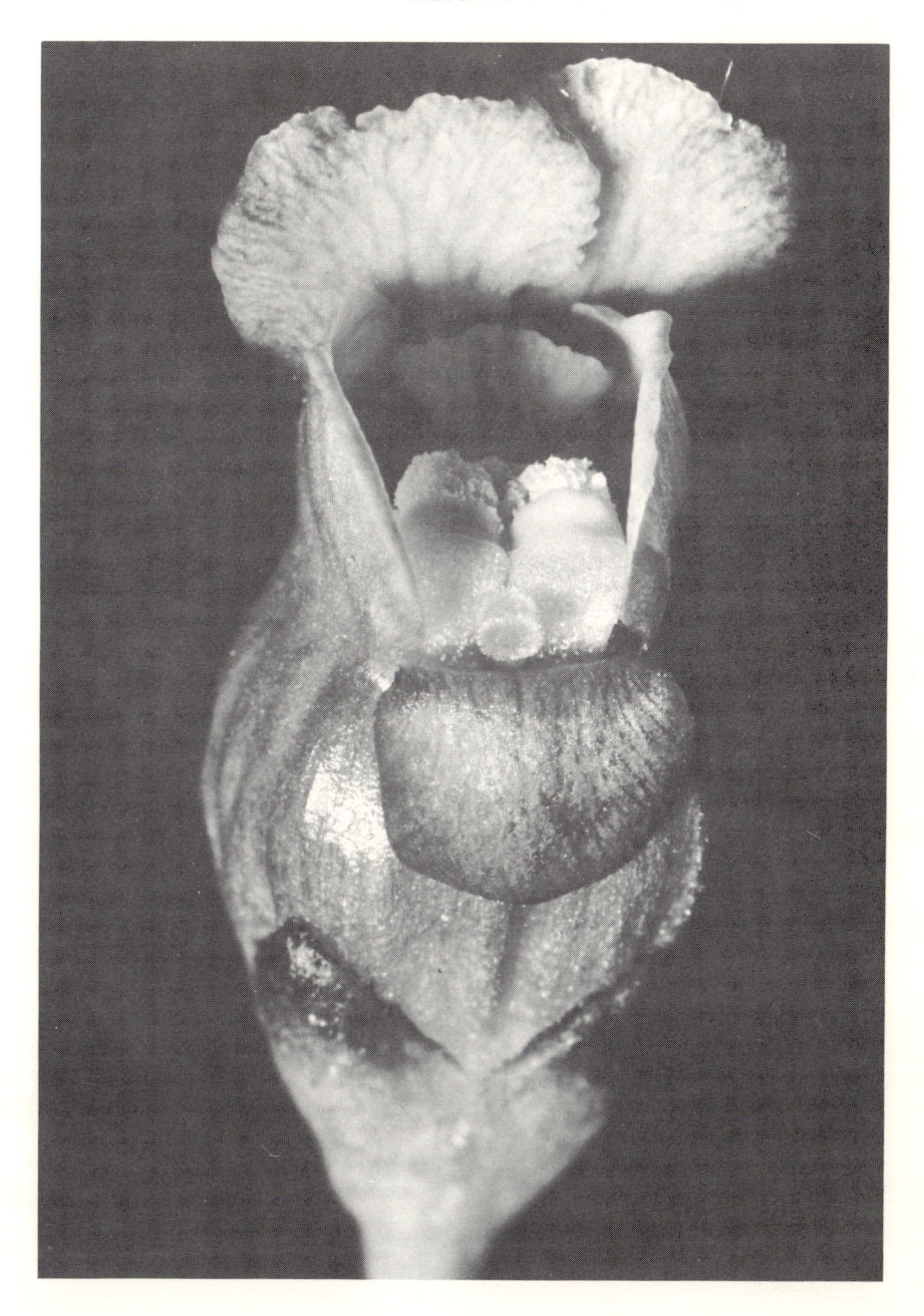

Figure 43. Figwort blossom.

NECTAR AND POLLEN PLANTS OF UTAH

and pollen from the blossoms in late summer.

Greasewood

Greasewood is an exceedingly abundant shrub in many places in Utah. It thrives on soils that carry too much salt for many other plant species. Frequently, greasewood is abundant in low poorly-drained stretches, while sagebrush appears on the hummock and on the hillsides surrounding the flats.

As a pollen source, it is worthy of special mention. The male blossoms open late in the afternoon and pollen showers from the blossoms without jarring, so that much is lost during the night. Many honey bees work the plants at close of day and a few are seen on them the next morning. Otherwise, but few bees are observed visiting the plants. Nectar secretion, if any, must be limited, because no nectar has been found on the plants (figure 44).

Figure 44. Greasewood is an important source of pollen in many places during June.

Gum plant

Several species of gum plant (*Grindelia* spp.) grow throughout the state in suitable locations. The plant thrives where overgrazing occurs, and in many places it is almost the only plant growing along roadside and ditchbanks, where cattle or sheep frequently pass. Apparently it is unpalatable and, as the other plants are eaten off time after time, this plant takes their places. The bright yellow blossoms frequently paint the roadsides for miles.

As a provider of nectar, gum plant is worthy of brief mention. Bees visiting the blossoms sometimes have a tiny amount of nectar visible in their stomachs. The nectar is yellow, and its presence in the hive un-

Figure 45. Gum plant, a valuable source of pollen.

NECTAR AND POLLEN PLANTS OF UTAH

doubtedly adds considerable color to the honey from plants that produce colorless nectar. The blossoms are open all day and each blossom apparently lasts several days, so that once this plant begins to blossom, it is continuously available (figures 45 and 46).

Bees work the plant freely for the pollen, which has a bright golden color. A reserve supply of this pollen has been observed in colonies at various places in northern Utah.

Honeysuckle

Honeysuckle (*Lonicera utahensis* S. Wats.), an attractive woody shrub, blooms in late spring and early summer with numerous yellow trumpet-shaped flowers in pairs. Nectar is secreted at the base of the corolla and lodged in the shallow cup. The plant blooms regularly

Figure 46. Close-up of gum plant blossom.

PLANTS

each year so that the nectar flow is certain. It is visited by many insects, the most important of which is the honey bee (figure 47).

They are associated with pine forests high in our mountains and where planted. They grow to about 5 feet tall.

Houndstongue

This introduced weed has been observed in many places in Utah. It grows on any land that has a moderate supply of water, and it is becoming abundant on foothill pastures in the vicinity of Richmond. The nectar, which is secreted freely, has a relatively high sugar content,

Figure 47. Honey bee collecting nectar from honeysuckle.

NECTAR AND POLLEN PLANTS OF UTAH

making it attractive to honey bees. It is reported to be of considerable importance at some locations as a fill-in between the dandelion and the beginning of alfalfa-sweet clover in mid-June. The seeds of this plant are covered with minute hooks, by which they cling tenaciously to clothing.

Linden or basswood

Although these plants are not native, both the American and German types now grow in numerous locations, particularly around homes and along city streets. Bees are attracted to both nectar and pollen, and procure large loads of the latter. The linden blossoms in midsummer (figures 48 and 49), and by that time the bees are usually strong enough to store honey. The nectar, as seen in the stomach of the bee, is colorless, and so is the honey when unmixed with that from other sources. Bees are frequently seen working

Figure 48. Linden blossoms yield much nectar as well as some pollen.

PLANTS

linden blossoms from early in the morning until almost dark. Although little surplus honey is produced from lindens in Utah, these trees contribute to the honey supply of midsummer.

Manzanita

In southwestern Utah, some of the hills are covered with this plant. It is a good source of colorless nectar. Honey bees collect its rather inaccessible gray pollen in the absence of better sources.

Maple

The trees blossom in spring and yield both nectar and pollen. Most of the native maple trees are in the mountains, where few honey bees

Figure 49. Close-up of linden blossom.

NECTAR AND POLLEN PLANTS OF UTAH

are now located. It is abundant in many locations. An illustration of the big-leaf maple is shown in figures 50 and 51.

Mesquite

This is a good source of white honey and cream-colored pollen. It is a heat lover and is found along the Arizona border.

Mustard

There are many species of mustard in Utah, but only a few are im-

Figure 50. Maple blossoms are copious providers of nectar and pollen in spring when bees need stimulation. Some mountainsides are covered with maple trees.

PLANTS

portant nectar and pollen sources. Black mustard [*Brassica nigra* (L.) Koch] introduced from Europe by Franciscan missionaries more than 100 years ago, is a good source of nectar and pollen. It blooms from late May to July and in some years a second crop will bloom from September until frost. Mustard pollen is very attractive to bees (figure 52).

Oaks

Various species of oak grow in Utah and are abundant in some places on the mountains and foothills. Gambel oak (*Quercus gambelii* Nutt.) is abundant on the lower western slopes of the Wasatch

Figure 51. Close-up of maple blossom.

NECTAR AND POLLEN PLANTS OF UTAH

Figure 52. Honey bee with load of mustard pollen collecting nectar.

PLANTS

Mountains. The predominant value of the oaks lies in their pollen production. They blossom late in the spring, providing a large amount of pollen, over a period of at least three weeks, to aid in building up the colony population.

Fortunately, bees taken into the oak sections are usually sufficiently removed from the agricultural and smelter areas to be fairly free from the danger of industrial poisoning. Many of the mountain canyons are suitable locations for apiaries, particularly before the time of the alfalfa honey crop.

The hook-covered blossoms yield both nectar and pollen. The plant is scattered over numerous places but does not persist in cultivated fields.

Onion

Onion (*Allium cepa* L. and spp.) is common in gardens and grown extensively in some areas for seed. Nectar is secreted by the ovary in the three notches between the carpels. The flowers are visited by many insects; the most important is the honey bee (figure 53).

The onion yields nectar freely where adequate moisture is available, and the honey is light amber in color with characteristic flavor which disappears after the honey is fully ripened and sealed.

Figure 53. Honey bee collecting nectar from onion blossoms.

NECTAR AND POLLEN PLANTS OF UTAH

Field poppy

Field poppy [*Roemeria refracta* (Steve.) DC.] was introduced from the Mediterranean area and has become a weed in wheat fields on the foothills. It grows about 12 inches tall. Its 2½-inch bright orange petals are jet black at the base with a narrow edge of white between the two colors with great masses of stamens. It blooms early in June and sometimes covers the land for miles with solid mass of bloom. Bees collect much pollen but little, if any, nectar (figure 54).

Prickly lettuce

A frequently occurring source of pollen during the morning hours. It is palatable to livestock and therefore largely confined to roadsides and waste places.

Rabbitbrush

The true rabbitbrush provides yellow nectar and orange-colored pol-

Figure 54. Honey bee collecting pollen from field poppy.

PLANTS

len. The blossoms continue even after moderate frosts. Honey is dark colored, strong flavored, and considered undesirable when mixed with alfalfa and clover honey (figures 55, 56, 57, and 58).

Safflower
(Carthamus tinctorius L.)

Grown for the flowers, which yield a dye, and for the seed as a source of oil.

Sagebrush
(Artemisia tridentata Nutt.)

Grows from 1 to 10 feet high with a stout trunk and a few ascending branches. The mature bark is gray and shreddy. It is rounded or somewhat flattened on top (figure 59). The evergreen leaves are gray-green to yellowish white and very aromatic. It grows on foothills and mountainsides in non-saline, rocky soil. Its blossoms are yellow and very small (figure 60). It blooms in late summer and fall. Bees collect much yellow pollen from the blossoms, especially when other pollen sources are scarce.

Sunflower

Sunflower (*Helianthus annuus* L.) is a tall annual herb up to 9 feet, with large composite heads several inches across; the disk flowers are brown and the large rays yellow (figure 71). It is one of the best known of American native flowers. The flowers are proterandrous; they are pollinated by bees and other insects. Bees visit the sunflower for nectar

Figure 55. Rabbitbrush, an important late summer source of nectar and pollen.

— 53 —

NECTAR AND POLLEN PLANTS OF UTAH

Figure 56. Rabbitbrush.

Figure 57. Close-up of rabbitbrush blossoms.

PLANTS

and pollen. Sometimes the plants are the source of large quantities of honey. The honey is yellow amber, with a strong flavor. It blooms from July to September.

Teasel

Teasel (*Dipsacus sylvestris* Huds.) is a stout, coarse perennial, 2 to 4 feet tall with prickly stems and toothed, prickly leaves, and blue flowers in dense heads. It is found in old fields, along ditch banks and roadsides. It blooms in early summer and produces a white honey with a good flavor. The pollen grains are white in color, large and densely covered with long spines (figure 62).

Waterleaf

Waterleaf (*Hydrophyllum capitatum* Dougl.) is a tender, crisp, erect herb that grows 4 to 12 inches tall. It can be found growing in damp, rich soil in open woods in the north-

Figure 58. Honey bee collecting nectar from rabbitbrush.

Figure 59. Sagebrush.

PLANTS

Figure 60. Flowers of sagebrush.

Figure 61. Sunflower blossom.

PLANTS

Figure 62. Teasel.

NECTAR AND POLLEN PLANTS OF UTAH

ern part of our state. It blooms in April and May with pale lavender or whitish blossoms in compact heads of 1 to 2 inches in diameter (figures 63 and 64). Apiaries located near woodland areas are often aided by this plant which secretes nectar freely at the base of the bell-shaped flower. Nectar can be seen in the cross section of the bell-shaped flower of *Phacelia* (figure 65).

Figure 63. Waterleaf.

— 60 —

PLANTS

Whitetop

This noxious perennial weed is thoroughly established in various places in the state. It blossoms early in the summer. The honey bee visits whitetop for pollen and, to a limited extent, for nectar.

Wild currant

Wild currants (*Ribes* spp.) are low shrubs, blossoming with racemes of golden flowers from the axils of the leaves (figures 66 and 67). As the blossoms mature, they become rose colored. The fruit is edible,

Figure 64.   Flowers of waterleaf.

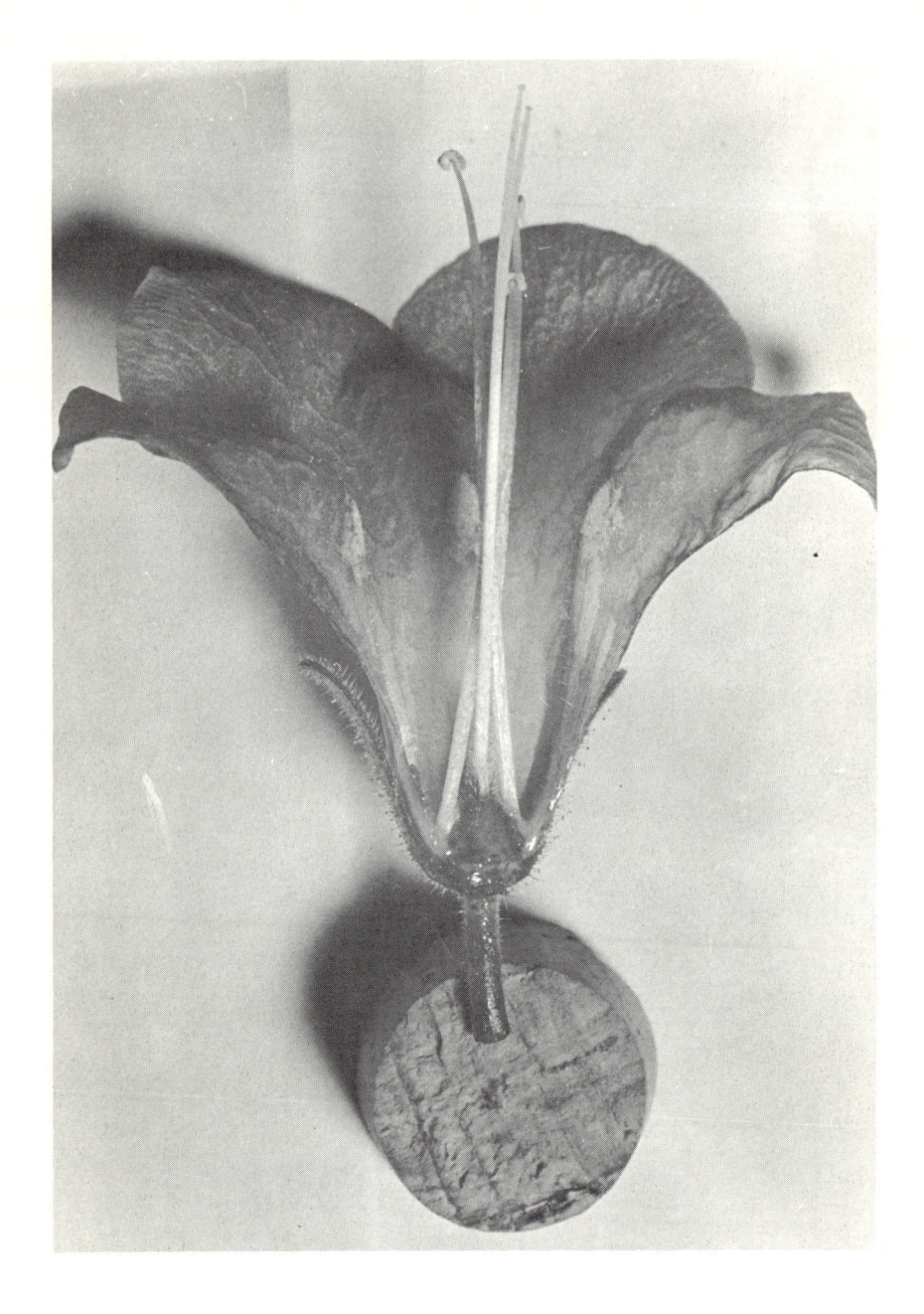

Figure 65. Bell-shaped flower showing nectar at base.

PLANTS

Figure 66. Wild currant blossoms.

NECTAR AND POLLEN PLANTS OF UTAH

sweet and may be any color ranging through yellow, red, and black. The nectar is free or sometimes concealed. It is secreted by the epigynous disk. Currants and gooseberries are dependent upon insect pollination and are well adapted to bees.

There is much variation in the insect relations of the genus *Ribes*. Some species have a long calyx tube and are pollinated by humming birds, though bees visit the flowers for pollen.

Figure 67. Close-up of wild current blossom.

PLANTS

Wild parsley
(Lomatium supp.) and carrot
(Daucus carota L.)

Common in early spring in our canyons and foothills. It provides an abundant source of pollen and some nectar needed for early spring brood rearing. Many insects visit carrot blossoms for nectar and pollen where they are grown for seed (figure 68).

Wild plum and chokecherry

These bush-like trees are covered with white blossoms in spring. Dense patches of wild plum grow along streams and ditch banks within reach of many apiaries in the valleys, while chokecherries grow mostly in the mountains where few apiaries are present. Bees visit them freely for nectar and pollen (figures 69 and 70).

Figure 68. Honey bee collecting pollen from carrot.

Figure 69. Chokecherry, found in canyons throughtout the state.

PLANTS

Willow

A dependable source of pollen and nectar. The female blossoms are inconspicuous but are visited freely by bees. The male catkins are conspicuous and familiar to all (figure 71).

CONCLUSIONS

The nectar in plant blossoms that is not harvested for honey by using the honey bee is an economic loss. Instead of being trespassers in alfalfa fields, gardens, and orchards, bees perform an indispensable service. During past years, honey production has been depended on to pay for the bees needed for both pollination and honey supply. In some sections, more bees are required for pollination than can be paid for with the income from the honey crop.

Figure 70. Close-up of chokecherry blossom.

NECTAR AND POLLEN PLANTS OF UTAH

Figure 71. Willow is an important source of nectar and pollen for stimulation of brood rearing in the spring.

Table 5. Nectar and pollen sources of Utah

Plant †	Blossoming period	Regions of occurrence	Usual habitat	Value as nectar source	Value as pollen source
Alder (**Alnus tenuifolia** Nutt.)	Mar., Apr.	1, 2	Borders of streams and mountain meadows	None	Minor
Alfalfa* (**Medicago sativa** L.)	June, Sept.	1, 2, 3, 4	Field	Major	Variable
Alsike clover (**Trifolium hybridum** L.)	June	1	Moist meadows	Minor	Minor
Apple* (**Malus pumila** Mill.)	April, May	1, 2, 4	Cultivated	Minor	Major
Apricot (**Prunus armeniaca** L.)	April, May	1, 4	Cultivated	Minor	Major
Arrowweed [**Pluchea sericea** (Nutt.) Cov.]	Spring	4	Ditch banks	Major	Minor
Asparagus (**Asparagus officinalis** L.)	Spring	1, 2, 3, 4	Cultivated and ditch	Minor	Minor
Aster (**Aster** spp.)	Fall	1	Moist waste places	Minor	Very little
Astragalus (**Astragalus** spp.)	Spring, summer	1, 2, 3, 4	Favorable spots	Some	Some
Baccharis (**Baccharis** spp.)	Spring, summer	4	Ditch banks	Minor	?
Balsamroot* [**Balsamorhiza sagittata** (Pursh) Nutt.] (Creeping)	Early summer	1, 2, 3	Open hillsides	Important	Important
Barberry (**Berberis repens** Lindl.)	Spring	1, 2	Semi-dry, gravelly hillsides, wider shrubs and trees	Some	Some

Table 5. (Continued)

Plant ‡	Blossoming period	Regions of occurrence	Usual habitat	Value as nectar source	Value as pollen source
Bassia [**Bassia hyssopifolia** (Pall.) Kuntze]	July-Aug.	3	General	None	Very little
Bitter brush* [**Purshia tridentata** (Pursh) DC.]	May-June	1, 2	Dry slopes	Some	Some
Birch (**Betula** spp.)	Mar., Apr.	1, 2	Along mountain streams	None	Minor
Blackberry (**Rubus** spp.)	May, June	1, 2	Where grown	Minor	Minor
Black locust (**Robinia pseudoacacia** L.)	June	1, 2, 3, 4	Where planted	Some	None
Blazing-star (**Mentzelia** spp.)	Summer	1	Gravel beds	?	Some
Beggar ticks (**Bidens** spp.)	Fall	1	Wet places	Minor	Some
Boxelder* [**Acer negundo** L., var. **interior** (Britt.) Sarg.]	April	1	Canyons, draws	?	Considerable
Broom snakeweed [**Gutierrezia sarothrae** (Pursh) Britton & Rusby]	July-Aug.	1, 2, 3, 4	Dry plains	Minor	Somewhat
Buffaloberry (**Shepherdia** spp.)	Early spring	1, 2	Moist places	Considerable	Important early
Burdock (**Arctium minus** Bernh.)	Summer	1	Old homesites	Little	Considerable
Buttercup, sagebrush* (**Ranunculus jovis**)	April, and June and July	1, 2	Foothills, high mountains beside melting snowdrifts	Some	Some

Table 5. (Continued)

Plant †	Blossoming period	Regions of occurrence	Usual habitat	Value as nectar source	Value as pollen source
Cactus* (**Optunia** spp.)	April-June	4	Stony slopes and gravelly washes	Some	Some
Canada thistle [**Cirsium arvense** (L.) Scop.]	Early summer	1	Where introduced	Some	Much
Catalpa* (**Catalpa speciosa** warder)	June	1, 2	Where introduced	Some	Some
Catclaw (**Acacia greggii** Gray)	Spring	4	Washes	Important	Important
Catnip (**Nepeta cataria** L.)	Summer	1	Old homesites	Some	?
Cattail (**Typha latifolia** L.)	Summer	1, 2, 3, 4	Wet places	None	Very little
Ceanothus (**Ceanothus** spp.)	Spring, summer	1, 2	Mountains	Variable	Considerable
Cheese weed (**Malva** spp.)	May-Sept.	1, 2	Widely distributed	Some	Some
Cherry Sour (**Prunus cerasus** L.) Sweet (**Prunus avium** L.)	Spring Spring	1 1	Cultivated Cultivated	Minor Minor	Minor Minor
Chicory (**Cichorium intybus** L.)	Summer, fall	1	Roadside	Some	Much
Chokecherry* [**Prunus virginiana** L., var. **melanocarpa** (A. Nels.) Sarg.]	May-June	1, 2	Mountains—along streams	Some	Important
Cleome* Rocky Mountain bee plant (**Cleome serrulata** Pursh)	July-Aug.	3	Favorable spots	Important	Some

Table 5. (Continued)

Plant †	Blossoming period	Regions of occurrence	Usual habitat	Value as nectar source	Value as pollen source
Clovers					
Red (Trifolium pratense L.)	June-Aug.	1, 2	Cultivated	Important	Important
Strawberry (Trifolium fragiferum L.)	June-Aug.	1, 2	Moist places	Minor	Minor
White (Trifolium repens L.)	June-Aug.	1, 2	Fields, waste	Important	Important
Corn (Zea mays L.)	July-Aug.	1, 2, 3, 4	Field, garden	None	Much
Cottonwood (Populus spp.)	Spring	1, 2, 3, 4	Moist places	?	Important early
Cucumber* (Cucumis sativus L.)	July-Aug.	1, 2, 3, 4	Cultivated	Some	Some
Currants (Ribes spp.)	Spring	1	Roadside, ditches	Some	Some
Dandelion* (Taraxacum officinale Weber)	Spring, summer	1	Meadows, lawns	Important	Important
Desert weed (Glycyrrhiza spp.)	Summer	1	Favorable spots	Some	Minor
Desert mallow (Sphaeralcea spp.)	Spring, summer	1, 2	Range	Considerable	Important
Elm* (Ulmus spp.)	Early spring	1, 2	Street, home	?	Considerable
Evening primrose* (Oenothera spp.)	May-July	1, 2, 3, 4	Favorable spots	Some	Some
Figwort* (Scrophularia lanceolata Pursh)	Summer	1, 2	Favorable spots	Some	Some
Filaree [Erodium cicitarium (L.) L'Her]	Spring	1	Open ground	Minor	Considerable

Table 5. (Continued)

Plant †	Blossoming period	Regions of occurrence	Usual habitat	Value as nectar source	Value as pollen source
Fireweed (**Epilobium angustifolium** L.)	July-Aug.	1, 2	Mountains	Minor	Minor
Globe-pod hoary cress [**Cardaria pubescens** (C. A. Meyer) Roll.]	Summer	1, 2	Waste places	Practically none	Little
Goat's rue or professor weed (**Galega officinalis** L.)	Summer	1	Where introduced	Some	Some
Golden rod (**Solidago** spp.)	July-Sept.	1, 2	Moist places	Minor	Minor
Grasses (**Poaceae**)	Summer	1, 2	Fields, meadows	None	Some
Greasewood* [**Sarcobatus vermiculatus** (Hook.) Torr.]	June-Aug.	1, 2, 3	Poorly drained soil	None	Important
Gum plant* [**Grindelia squarrosa** (Pursh) Dunal]	Summer, Fall	1, 2	Roadside, range	Minor	Variable
Hawksbeard (**Crepis** spp.)	May-Aug.	1, 2	Meadows and dry hillsides	Some	Some
Hawthorn (**Crateagus douglasii** Lindl.)	Late April and May	1	Stream banks	Some	Some
Hemlock, poison (**Conium maculatum** L.)	Summer	1	Waste places	Some	Some
Hoarhound (**Marrubium vulgare** L.)	Summer	1	Sheep trails	Important	None
Honey locust (**Gleditsia triancanthos** L.)	July	1	Where planted	Minor	Some
Honeysuckle* (**Lonicera utahensis** S. wats)	May-June	1, 2	Mountains and where planted	Some	Some

— 73 —

Table 5. (Continued)

Plant †	Blossoming period	Regions of occurrence	Usual habitat	Value as nectar source	Value as pollen source
Houndstongue (**Cynoglossum officinale** L.)	May-June	1	Pasture	Important	Minor
Horsechestnut (**Aesculus hippocastanum** L.)	May-June	1	Ornamental	Some	Some
Horsemint [**Agastache urticifolia** (Benth.) Kuntze]	July-Aug.	1, 2, 3	Canyons, wet and dry situations	Minor	Minor
Ladino clover (**Trifolium repens** L.)	May-Sept.	1, 2	Irrigated pasture	Important	Some
Lilac (**Syringa** spp.)	Spring	1	Where planted	Some	Some
Linden* or basswood (**Tilia** spp.)	July	1	Where planted	Important	Minor
Lupine (**Lupinus** spp.)	Spring, summer	1, 2, 3, 4	Roadsides	?	Minor
Manzanita (**Arctostaphylos** spp.)	Spring	4	Mountains	Important	Minor
Maple* (Big tooth) (**Acer grandidentatum** Nutt.)	Late spring	1, 2	Mountains	Important	Important
Norway and other spp. (**Acer** spp.)	Late spring	1	Ornamental	Some	Some
Mesquite [**Prosopis julifora** (SW.) DC.]	Early summer	4	Washes	Important	Important
Milkweed (**Asclepias** spp.)	Summer	1, 2, 3, 4	Miscellaneous	Variable	None
Mountain mahogany (**Cercocarpus** spp.)	Early summer	1, 2	Mountains	?	Some
Mule ears (**Wyethia amplexicaulis** Nutt.)	Early summer	1, 2, 3	Open hills	Important	Important

— 74 —

Table 5. (Continued)

Plant †	Blossoming period	Regions of occurrence	Usual habitat	Value as nectar source	Value as pollen source
Mustards* (Brassica spp. and sisymbrium spp.)	Spring, summer	1, 2	Fields, roadside	Variable	Major
Oak (Quercus spp.)	Late spring	1, 2, 3	Mountains	None	Important
Onion* (Allium spp.)	Summer	1, 4	Seed field	Considerable	Very little
Orchard morning-glory, or bindweed (Convolvulus arvensis L.)	Summer	1	Where introduced	Some	Considerable
Peach* [Prunus persica (L.) Batsch]	Spring	1	Cultivated	Important	Important
Peas (Pisum sativum L.)	Early summer	1	Cultivated	None	Little
Pecans (Carya illinoensis Koch)	Early spring	4	Cultivated	Some	Some
Penstemon (Penstemon spp.)	May-July	1, 2	Hillsides, mountain slopes	Some	Some
Pepper grass (Lepidium spp.)	Summer	1, 2	Waste places	Some	Some
Pickleweed (Salicornia utahensis Tides.)	Late summer	3, 1	Wet places	Some	Considerable
Pine (Pinus spp.)	May-June	1, 2	Mountains—cultivated	None	Some
Plantain or ribwort (Plantago lanceolata L.)	Summer	1	Pasture	None	Much
Poison oak (Rhus radicans L.)	Early summer	1	Ditch, roadside	Minor	Minor
Pomegranate (Punica granatum L.)	Summer	4	About homes	Some	Some

Table 5. (Continued)

Plant ‡	Blossoming period	Regions of occurrence	Usual habitat	Value as nectar source	Value as pollen source
Poppy, field* [**Roemeria refracta** (Steve.) DC.]	Early summer	1	Hillsides and wheat fields	None	Minor
Saltbrush (**Atriplex** spp.)	June-Sept.	1, 3	Range	None?	Some
Seepweed (**Suaeda** spp.)	Summer	1, 3	Wet places	None	Considerable
Segolily (**Calochortus nuttallii** Torr.)	Early summer	1	Favorable spots	Minor	Minor
Service Berry [**Amelanchier alnifolia** (Nutt.) Nutt.]	April-May	1	Mountains—dry soil on hillsides	Minor	Minor
Smartweed (**Polygonum** spp.)	Fall	1	Wet places	Some	?
Snowbury [**Symphoricarpos oreophilus** Gray var. **utahensis** (Rybd.) A. Nels.]	May-June	1, 2	Gardens and mountains	Minor	Minor
Spearmint (**Mentha spicata** L.)	Summer	1	Wet places	Important	?
Squawbrush (**Rhus trilobata** Nutt.)	April-May	1, 4	Dry hills and plains	Important	?
Stickseed [**Hackelia patens** (Nutt.) I. M. Johnst.]	May-June	1, 2	Foothills	Some	Minor
Strawberry (**Fragaria chiloensis** Duchesne, var. **ananassa** Bailey)	May	1, 4	Where planted	Some	Some
Sugar beet seed (**Beta vulgaris** L.)	Summer	1, 4	Where grown	Very minor	Very minor
Sumac (**Rhus** spp.)	Summer	1	Where placed	Some	Some

Table 5. (Continued)

Plant †	Blossoming period	Regions of occurrence	Usual habitat	Value as nectar source	Value as pollen source
Little sunflower [Helianthella uniflora (Nutt.) Torr. & Gray]	May-Aug.	1, 2	Dry canyon hillsides	Major	Major
Sunflower common* (Helianthus annuus L.)	Summer, fall	1, 2	Field, roadside	Minor	Minor
Sunflower, Nuttall's (Helianthus nuttallii Torr. & Gray)	Aug.-Nov.	1, 2	Streambanks and moist places	Minor	Minor
Sweet clover*					
White (Melilotus alba L.)	May-Nov.	1, 2, 3, 4	Pasture, roadside	Major	Major
Yellow [Melilotus officinalis (L.) Lam.]	May-Sept.	1, 2, 3, 4	Pasture, roadside	Major	Major
Tamarix (Tamarix pentandra Pall.)	Summer	1, 2, 3, 4	Cultivated and escaped moist places	Some	Some
Teasel* (Dipsacus sylvestris Huds.)	Summer	1	Roadside	Potentially important	Some
Thistles (Cirsuim spp.)	May-Sept.	1	Mountain meadows and dry rocky slopes	Minor	Minor
Timothy (Phleum pratense L.)	Early summer	1	Moist places	None	Considerable
Tree-of-Heaven [Ailanthus altissima (Swingle)]	May-June	1	Ornamental	Some	Some
Water leaf* (Hydrophyllum capitatum Dougl.)	April-May	1	Damp, rich soil in open woods	Minor	Minor

— 77 —

Table 5. (Continued)

Plant †	Blossoming period	Regions of occurrence	Usual habitat	Value as nectar source	Value as pollen source
White clover* (**Trifolium repens** L.)	May-Sept.	1, 2	Irrigated pasture	Important	Some
White sweet clover* (**Melilotus alba** Desv.)	June-Sept.	1, 2	Pasture, roadside	Major	Major
Whitetop [**Cardaria draba** (L.) Desv.]	Early summer	1, 3	Where introduced	Very minor	Very minor
Wild carrot* (**Daucus carota** L.)	Early summer	1	Waste places	Minor	Minor
Wild currant* (**Ribes** spp.)	Early summer	1, 2	Favorable spots	Some	Some
Wild geranium (**Geranium** spp.)	Early summer	1	Roadside	Considerable	Considerable
Wild lettuce (**Lactuca serriola** L.)	Summer, fall	1, 2	Roadside	None	Slight
Wild parsnip (**Pastinaca sativa** L.)	Summer	1, 2	Moist places	Some	Some
Wild plum (**Prunus americana** Marsh)	Late spring	1, 2	Mountains	Some	Important
Wild rose (**Rosa** spp.)	Early summer	1, 2	Moist places	Some	Considerable
Willow* (**Salix** spp.)	Spring	1	Ditch banks	Important	Major
Yellow sweet clover [**Melilotus officinalis** (L.) Lam.]	June-Sept.	1, 2	Pasture, moist places	Major	Major
Yerba-santa or mountain balm (**Eriodictyon angustifolium** Nutt.)	Early summer	4	Favorable spots	Important	Little

LITERATURE CITED

Andersen, Berniece A., and Arthur H. Holmgren. 1966. Mountain plants of northeastern Utah. Utah State Ext. Serv. Circ. 319.

Judd, D. B. 1940. Hue saturation and surface colors (and Rev. Sci. Instruments). J. Opt. Soc. Amer. 30:2.

Merz, A., and Paul M. Rea. 1950. A dictionary of color. 2nd ed. McGraw-Hilli, New York, New York. 208 p.

Oertel, E. 1967. Nectar and pollen plants. *In* Beekeeping in the United States, Agriculture Handbook No. 335. Agr. Res. Serv., USDA, Govt. Printing Office, Washington, D.C., p. 10-16.

Percival, Mary. 1947. Pollen collection by *Apis mellifera*. New Phytol. 46:142-173.

Standifer, L. N. 1947. Honey bee nutrition. *In* Beekeeping in the United States, Agriculture Handbook No. 335. Agr. Res. Serv., USDA, Govt. Printing Office, Washington, D.C., p. 52-55.

Todd, Frank E., and Ormond Bretherick. 1942. The composition of pollens. J. Econ. Entomol. 35: 312-217.

Vansell, G. G. 1931. Nectar and pollen plants of California. Calif. Agr. Exp. Sta. Bull. 517, 55 p.

Vansell, G. H. 1949. Pollen and nectar plants of Utah. Utah Agr. Exp. Sta. Circ. 124. 28 p.

Webster's New International Dictionary of the English Language. 1947. G. & C. Merriam Co., Springfield, Massachusetts.

Wilson, W. T., J. O. Moffett, and H. D. Harrington. 1958. Nectar and pollen plants of Colorado. Colo. Agr. Exp. Sta. Bull. 503-S. 72 p.

INDEX

— A —

Acer negundo, 24
Agastache urticifolia, 13, 14
alfalfa, 9, 16, 17, 18, 19, 20
Allium cepa, 51
almond, *see* deciduous fruits
alsike clover, *see* clover
apple, *see* deciduous fruits; rosaceous type pollen
apricot, *see* deciduous fruits
Arctostaphylos, see manzanita
Artemisia tridentata, 53, 56, 57
aster, 10, 11, 20
Astragalus, 21
   A. lentiginosus, 21

— B —

Balsamorhiza, 21
   B. sagittata, 11
balsamroot, 11, 21
basswood, 10, 46
beekeeping regions, 1, 2, 3, 4
bitterbrush, 22, 23
black locust, 24
black mustard, *see* mustard
boxelder, 24
Brassica, 11, 12
   B. nigra, 49, 50
broom snakeweed, 24
buttercup, 25

— C —

cactus, 10, 11, 25, 26
cane fruits, 25
carrot, wild, *see* wild carrot
Cardaria draba, *see* whitetop
Carthamus tinctorius, 53
Catalpa, 26, 27, 28
   C. speciosa, 26, 27, 28
catnip, 13, 14
cheeses, 14
chemical analysis of pollen, 14, 15, 16
cherry, *see* deciduous fruits
chokecherry, 65, 66, 67
Cichorium intybus, 11
Cleome, 26, 29
clover, 9, 10, 28, 29, 30, 31
composite type pollen, 10, 11
coniferous type pollen, 11, 12
cottonwood, 31
cucumber, 31
cruciferous type pollen, 11, 12
Cucumis sativus, 31
currant, *see* wild currant

— D —

dandelion, 11, 12, 31, 32, 33
Daucus carota, 14, 65
deciduous fruits, 31, 34, 35, 36, 37
Dipsacus syvestris, 55, 59
Dutch clover, *see* clover

— E —

elm, 38
evening primrose, 11, 13, 38, 39, 40

— F —

field poppy, 52
figwort, 39, 41

— G —

Gambel oak, *see* oak
grass, 12, 13
greasewood, 42
Grindelia, 43, 44
gum plant, 43, 44

— H —

Helianthus, *see* sunflower
honeysuckle, 44, 45
horehound, 13, 14
horsemint, 13, 14
houndstongue, 45
Hydrophyllum capitatum, 55, 60, 61, 62

— I —

Iliamna rivularis, 14

— L —

labiate type pollen, 12, 14, 15
leguminous type pollen, 9, 10
lettuce, *see* prickly lettuce
linden, 10, 46, 47
loco weed, 21

INDEX (Continued)

locust, *see* black locust
Lonicera utahensis, 44, 45

— M —

mallow type pollen, 14
Malus, 14
Malva neglecta, 14
manzanita, 47
maple, 47, 48, 49
Marrubium vulgare, 13, 14
matchweed, *see* bloom snakeweed
Medicago sativa, see alfalfa
Mentha spicata, 13, 14
mesquite, 48
mule ears, 11
mustard, 11, 12, 48, 49, 50

— N —

nectar secretion, 4, 5
Nepeta cataria, 13, 14

— O —

oak, 49
Oenothera, 11, 13
   O. caespitosa, 38, 39, 40
onion, 51
Opuntia, 10

— P —

parsley, *see* wild parsley
peach, *see* deciduous fruits
pear, *see* deciduous fruits
Phacelia, 60, 62
plum, *see* deciduous fruits; wild plum
poisonous plants, 4
poppy, field, 52
Populus, 31
prickly lettuce, 52
primrose, evening, *see* evening primrose
Prosopis julifora, see mesquite
Purshia tridentata, 22, 23

— Q —

Quercus gambelii, 49

— R —

rabbitbrush, 52, 53, 54, 55
Ranunculus jovis, 25
raspberries, *see* cane fruit
red clover, *see* clover
Ribes, 61, 63, 64
Robinia pseudo-acacia, 24
Rocky Mountain bee plant, 26, 29
Roemeria refracta, 52
rosaceous pollen type, 14

— S —

safflower, 53
sagebrush, 53, 56, 57
Salix, see willow
Sarcobatus vermiculatus, see greasewood
scarlet globemallow, 14
Scrophularia lanceolata, 39, 41
small rabbitbrush, *see* broom snakeweed
spearmint, 13, 14
Sphaeralcea munroana, 14
sunflower, 53, 58
sweet clover, *see* clover

— T —

Taraxacum officinale, 11
teasel, 55, 59
Tilia, 10
Trifolium, see clover

— U —

Ulmus, see elm
Umbelliferous type pollen, 14

— W —

waterleaf, 55, 60, 61, 62
white clover, *see* clover
whitetop, 61
wild carrot, 14, 65
wild currant, 61, 63, 64
wild hollyhock, 14
wild parsley, 65
wild plum, 65; *see* also deciduous fruits
willow, 67, 68
Wyethia, 21
   W. amplexicaulis, 11